APM – ACostE Cost Estimating

Association for Project Management

Association for Project Management
Ibis House, Regent Park
Summerleys Road, Princes Risborough
Buckinghamshire
HP27 9LE

The Association of Cost Engineers
Administration Office
Lea House
5 Middlewich Road
Sandbach
Cheshire
CW11 1XL

First edition 2019

British Library Cataloguing in Publication Data is available.
Paperback ISBN: 978-1-903494-84-4
eISBN: 978-1-903494-85-1

Cover design by Fountainhead Creative Consultants
Typeset by RefineCatch Limited, Bungay, Suffolk
in 10/14pt Foundry Sans

Contents

Table of Figures

Foreword

Estimation and its diligent application within any project is one of the cornerstones of successful project delivery. This guide has been created to assist the current and next generation of project stakeholders to understand the core values, information sets and underpinning knowledge that, if applied diligently, will improve the clarity and robustness of an estimate, informing the organisation with transparency and clarity and supporting better decision making.

The Association for Project Management (APM) and the Association of Cost Engineers (ACostE) have collaborated to bring this guide to you. All the collaborators have an excellent understanding of the challenges faced when generating an estimate that is fit for purpose, having had the experience of real-life situations where good estimates made the difference in delivering the project, and a burning desire to share their combined knowledge for the benefit of all.

Professor Andy Langridge
Director of business development, ARES Corporation

Acknowledgements

The development of this guide was a time-consuming and intellectually stimulating activity, which required the dedication of a few individuals who gave their time freely. It is important these people and organisations are recognised as a way of saying thank you.

Stephen Jones BEng(Hons) MSc MIET RPP FAPM, Sellafield Ltd
Sandy Cordon MSc CPFA, Audit manager, National Audit Office
Alan Jones BSc (Hons) ARCS MBA MMS FACostE CCE/A CCE ICECA, Estimata Limited
Andrew Nolan, Rolls-Royce
Tony Purpouri, CEng MEng(Hons) MRAeS CCE/A, IMD Group

Thank you to BAE Systems PLC, Rolls-Royce PLC, and Sellafield Ltd for granting permission to use their copyrighted material and/or resources and facilities in the production of this guide.

Additional material and comments were provided by:

Dale Shermon, BA, CDip (A&F), CCEA-P, CEng MIET, FACostE, MAPM, QinetiQ Fellow, Managing consultant
John Millington, Head of estimating, Sellafield Ltd
Stephanie Baker MACostE CPCostE ICECA, Principal cost engineer, DE&S MOD

Purpose and reason for this guide

The purpose of this guide is to provide a fundamental understanding of the methods of cost estimating and it explains a number of standard approaches available to promote good practice.

Estimates are critical to the project manager as they are needed to make informed decisions about projects across the different stages of the whole project life cycle, hence the cooperation of the Association of Cost Engineers (ACostE) and the Association for Project Management (APM) in publishing this guide. In project management, effective monitoring of a project's performance depends on having an appropriate, high-quality estimate against which progress can be measured.

For instance:

- Investment committees require estimates in order to predict return on investment and hence determine whether to support project proposals and the level of finance to invest in them.
- Cost estimates form part of option appraisals.
- Mature organisations often review projects at key stages and require them to meet internal governance guidelines on cost-benefit-risk.
- Organisations that manage a portfolio of projects require credible cost estimates in order to manage the spending profile of the portfolio.
- Good management practice needs to understand the impact of a project change, including risks and opportunities.

There is long-standing evidence that underestimation of project costs is a key reason for project failure. For instance, the National Audit Office's 2001 report *Modernising Construction* (Comptroller and Auditor General, 2001) found that limited understanding of the true cost was one of the main reasons that 70 per cent of public sector construction projects were delivered late or over budget. Industry surveys (KPMG, 2015) indicate that this is still the case. Academics have collected evidence for overruns costing billions of pounds in major infrastructure projects worldwide (Flyvbjerg, 2003). More recent studies (Comptroller and Auditor General, January 2017) have shown that these errors were often due to not taking estimating seriously enough, hampered by poor quality data and unrealistic assumptions (National Audit Office, December 2013).

By their very nature estimates are speculative; the word estimating is synonymous with approximation and guessing, yet estimates are vital for sound decision-making, planning and financial management.

Different techniques may be appropriate at different stages in a project's development. This guide will focus on cost estimating method and approaches. However, the advice in this guide is not limited to initial cost estimates; it is equally applicable to forecasting and to other forms of estimating, for example, time, schedule or performance.

The guide is not limited to public sector construction, infrastructure or defence projects: it is equally applicable to the private sector and a wider range of projects, for example, if you are recruiting a new member of staff, or building the world's fastest car, launching a new service, or licensing a new drug – every project needs a robust estimate.

Estimating framework

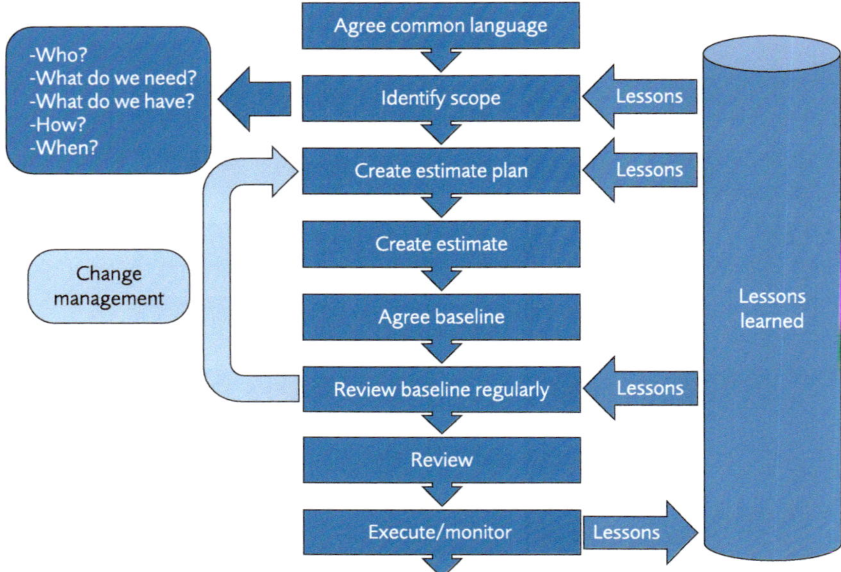

Figure 0.1 Estimating framework

Estimating consists of a number of activities, which provide a framework for generating and continuously improving an estimate. The diagram above shows a typical estimating framework, which includes the activities covered in the guide.

A definition of the terms used in the diagram above can be found in the glossary and are explained throughout this guide.

1

Developing the estimating plan

This section sets out what needs to be in place to construct a valid cost estimate: an understanding of who it is for, its scope, the information required and who needs to be involved. The degree of planning required will be scalable and flexible depending on the purpose of the estimate and the likely value or the strategic importance of the project, e.g. multiple conceptual studies for comparison purposes will not be planned with the same level of detail as an estimate committing multi-billions of pounds.

Estimates are based on whatever data is available, produced by using a range of tools, techniques and expertise. Estimating is an iterative process and estimates should be refined throughout the project life cycle as information matures and evolves. See Figure 0.1: Estimating framework.

1.1 Stakeholder engagement and mobilisation

1.1.1 Stakeholder commitment

The first activity in creating any cost estimate is to identify and engage with key stakeholders of the estimate to define their interests and roles. Roles are often identified using the RACI methodology, which classifies them as responsible, accountable, consult and inform. This will normally include senior management, the estimate owner or other people of influence over the estimating requirements. There are significant benefits to engaging with this group early, such as ensuring there is enough senior commitment to the process and requirements, and gaining an early understanding of the true objectives of the estimate to plan the estimating strategy.

1.1.2 Responsibility assignment matrix (RAM)

The responsibilities for the work can be documented using a responsibility assignment matrix (RAM) which is a "diagram or chart showing assigned responsibilities for elements of work" – (APM, 2019). For estimating, the focus should be on what the roles and responsibilities are to support the estimating process and who is to provide the appropriate information.

The estimating RAM will be an extension of the organisation or project RAM. It is a key document that needs to be collated or defined as early as possible, and needs to consider the estimating activities and support of estimating specific to the cost estimate. The RAM will identify all stakeholders who are considered responsible or accountable. These stakeholders may include technical leads, engineers, project managers, schedulers, risk managers, finance, commercial, and senior organisation and project leaders. Essentially, this means anyone impacted by the estimate, who has any data, information, interest or influence over the estimate.

1.1.3 Interpretation of stakeholder objectives and targets

One of the benefits of engaging early with key stakeholders is to understand the overarching objectives of the estimate, including its intended use and the audience to which the estimate will be presented. In addition, this is an opportunity to identify key dates, resource limitations, operating assumptions and other influences that may limit or assist the estimating process. For example, it enables the estimator to test the knowledge, understanding and interpretation of the intended requirements with key stakeholders.

1.2 Understanding the estimate scope

1.2.1 Scoping the estimate

The scope of the estimate can be determined using a blend of methods appropriate for the estimate purpose. Initially a product, service, work or organisational breakdown structure (PBS/SBS/WBS/OBS) can be used to understand the broad scope of the estimate. Depending on the purpose for which the estimate is required, requirements definition documents can be used to understand the scope. In conjunction with stakeholder validation, a picture can be compiled describing the purpose of the estimate and how best to approach its creation. For a simple estimate, this pictorial view may be sufficient. However, for more complex projects or products, with many interfaces and dependencies, a more formalised method may be required. This could include systems engineering and enterprise architecture methods, both of which are powerful tools to help define the scope. The drawback with these methods is that they are labour intensive so need to be tailored to the size and complexity of the estimate.

In addition to the basic scope it is essential to understand and capture the assumptions that underpin it, any internal or external dependencies, risks, opportunities and exclusions. This can be referred to as ADORE, (Shermon D, 2017).

Assumptions: The assumptions underpinning the estimate need to be determined up front. These will be documented, which will help to clarify areas of uncertainty in the scope of the estimate. Assumptions are statements that are taken to be true for the purpose of the estimate but may be unknown in reality, and are used to bound uncertainty in the estimate. For example, the level of future escalation may be unknown, but an agreed rate of escalation might be used. The project or organisation should have a method of capturing assumptions; there will usually be a master data assumptions list (MDAL) or a cost data and assumptions list (CDAL). These need to be closely managed and maintained within the wider assumptions management process.

Dependencies: These are a special class of assumptions that require prior information or an activity to have been completed in order for the estimate to be valid. For example, if an activity is scheduled to be completed on a specific date, and requires delivery of a product or information, in an agreed condition, then failure to meet this requirement may affect the cost and/or schedule.

Dependencies can take many forms and a structured approach is needed to ensure no key issues are missed. Frameworks such as PESTLE (political, economic, social, technological, legal and environmental) are used, but dependencies can often include the transactional supply of information or goods. These may apply to a greater or lesser extent depending on the estimate purpose or scope. Once the dependencies have been identified, they can be treated as interfaces and managed accordingly. It is important to define what impact these dependencies have on the estimate and process, including whether the cost of these interfaces should be captured within the estimate.

Particular consideration should be given to external dependencies between the contractor and the client/customer organisations, as well as those situations outside the control of both organisations, but which impact on the validity of the cost estimate assumptions. In these circumstances, it is advisable that there are appropriate contractual arrangements that define the responsibilities and expectations, and any recourse mechanisms.

If the dependencies are internal to the contractor organisation, or within their sphere of control through the supply chain, then such dependencies might be better considered as risks or opportunities to the project. Their potential effects might then be reflected in the schedule critical path and any consequential impacts evaluated in terms of cost and schedule performance.

Opportunities and Risks: A risk is something that may or may not occur but if it does occur, it is detrimental to the project. An opportunity is a positive risk; i.e. if it does occur it is beneficial to the project. NB: some organisations define a risk as negative, whereas others see it as positive or negative, and refer to a negative risk as a threat.

There are two types of risks and opportunities that need to be considered: those that impact on the project, and those that impact on the development of the estimate. The latter is concerned with the risks and opportunities of conducting the estimating process, including the assumed scope of work. As an example, a risk in this area could mean that a team is incorrectly sized for the level of work expected.

Exclusions: These are a specific type of assumption which state that the activity or event are out of the project's scope for the purpose of the estimate.

Any exclusions to the estimate are just as important as the inclusions and dependencies. Exclusion to the estimate could be made for a variety of reasons, including accounting, commercial or technical reasons. For each exclusion, it is important to identify the authority for that exclusion and ensure that they understand the reasons for the decision.

The scope of the estimating, including the ADORE list, should be shared and discussed with all stakeholders, giving them the opportunity to review and comment before the estimate is generated. This should avoid criticism of the cost estimate when completed.

1.2.2 Understanding the estimate maturity requirements

Maturity refers to the robustness of the estimating practices used to develop the estimate, including the data sources, tools, people and processes employed. In general, mature estimates are expected to be more accurate than immature estimates, although the robustness of the data can affect the precision. Precision is expressed using three points: the minimum, most likely, and maximum; this is known as a 3-point estimate. The term minimum does not necessarily refer to the absolute minimum value, but a credible (evidence-based) minimum – i.e. the true value is unlikely to be less than the stated minimum. The same applies for maximum, but at the other end of the distribution.

The maturity requirement of an estimate is predominantly driven by the intended use of the estimate. If the estimate is needed for a budget, then human nature is to look for a high level of maturity early in the project. In reality, for a pre-concept rough order of magnitude (ROM) costing, the maturity can be quite low. A 3-point estimate (3PE) is an indication of the level of uncertainty, error, noise or variability in the historical data upon which the estimate is based. The maturity of an estimate is an indication of the level of sophistication or confidence that can be placed in the estimate as a result of the data, tools, people and process used to generate it. Although unusual, it is possible to have a very narrow range 3PE with very limited maturity.

Resourcing limits, time constraints or lack of data availability often mean a higher maturity estimate is not possible, or is more difficult to achieve. These limits need to be communicated to senior stakeholders early to set expectations and manage these risks.

There are many ways in which the maturity of an estimate can be measured, reflecting the perception of the confidence one can have in the estimate. On a scale of one to nine, a maturity level of one may indicate an expert's opinion, whereas a nine represents a "make to print[1]". Clearly an expert opinion will have significant uncertainty compared with an actual spend history.

The purpose of the estimate will drive the requirement for a minimum maturity level; a budgetary estimate will typically need a higher maturity than a concept phase investment appraisal. This will drive timescales to create the estimate at the specified maturity level as well as resourcing requirements.

[1] Make-to-print means ready to place order with a supplier purely for manufacture having completed the design, development and testing

1.3 Manage the information requirements

1.3.1 Identifying the information requirements

There are many sources of information of varying applicability and trustworthiness. The location and availability of such information can vary significantly. A review with stakeholders and subject matter experts (SMEs) is the first step in identifying what data exists and from where it may be obtained. Existing estimates and models are a good source of information, although their maturity may be lower than required. In the absence of better information, existing estimates can be a good start by signposting the estimator in the direction of more relevant data should it become available.

If no previous estimates exist then the estimator should spend appropriate time sourcing high-quality data and assumptions from many locations. These could be from suppliers, public domain or internal stakeholders (e.g. finance/ commercial). This provisional data can be used as long as its limitations and assumptions are recorded and understood.

Having identified the information source, the next step is to freeze the data under configuration control.

There are numerous factors that need to be considered when collating and storing the information, i.e. its format, whether explicit permission is required to access and use the information, whether it can be retrieved easily, and whether there are any special handling requirements. Classified, commercial and personal information all need to be treated according to their appropriate legislative requirements.

Security issues will affect who can access the data and where they can use it, while commercial information may be subject to a non-disclosure agreement or intellectual property rights. Personal information must be handled in accordance with the Data Protection Act (legislation.gov.uk, 2015) and EU General Data Protection Regulation (GDPR). All data collated must be under permission of the data owner, where appropriate. This may also affect how the final outputs are presented and to whom.

In addition to where it is stored, the transmission media must adhere to the restrictions on the information.

Such data restrictions could limit the level of detail recorded in the basis of estimate (section 2.4).

1.3.2 Storing information

Having collated the information, it is essential that the estimating team can retrieve it. A logical method of recording the source of the data, as well as handling instructions, is essential. A simple log should be made of:

- what data is received;
- who received it;
- who sent it;
- what format it is in;
- permission and handling instructions;
- the date;
- file name, using a consistent and logical naming convention;
- any other pertinent information (emails, meeting minutes etc.);
- adequate version control to prevent loss of original information.

This will enable effective and compliant storing and retrieval of information.

In the case of long and complex programmes or large organisations, software obsolescence may hinder information retrieval in the future. If data is in such a format, consideration should be given as to whether it is appropriate to migrate the data to a more accessible format.

Back-up of the information is essential to ensure continuity in the event of data loss.

1.3.3 Reviewing the maturity of the information available

As will be discussed later, the maturity of the data will have a direct impact on the maturity of the estimate produced. Using the maturity grading described in this guide, the information should be assessed for its individual maturity scoring. This will take into account the source of the information, its fidelity, age and applicability to the project or organisation. It is critical that each piece of information is assessed independently and consistently to avoid any bias in the maturity scoring. The scored maturity and rationale must be recorded with other metadata relating to the information.

1.3.4 Revising the ADORE in line with information maturity risk

Once all data has been collated, stored, recorded and assessed, the assumptions, dependencies, opportunities, risks and exclusions (ADORE) need to be reviewed and revised to account for the individual and collective data quality, completeness and maturity. At this point it is likely that a significant amount of information is being collated, and so an individual register for each ADORE area may be necessary to record the information at a lower level than previously required.

1.3.5 Data normalisation

Data normalisation is the act of adjusting or categorising data to achieve a state whereby the normalised data can be used for comparative purposes in estimating. Data normalisation may be required for any of the following reasons:

- differences in work content or complexity (e.g. 20 per cent more work content);
- effects of inflation or escalation (e.g. 2.3 per cent annual escalation over the last three years);
- changes in accounting policies (e.g. changes in definitions of those tasks which attract direct versus indirect hours or costs);
- alternative measurement systems or scales (e.g. imperial versus metric scales);
- learning from experience (e.g. learning curves, productivity improvements).

As data normalisation is an essential consideration for any estimating method selected, it should be read as implicit within this guide that data normalisation is an integral step in all the estimating methods outlined in section 2.2.

The equivalent of a data normalisation procedure may also be applied as the last step of an estimating method in order to convert the estimate derived for some base reference conditions into some future (or alternative past) outturn reference point for internal or external stakeholders.

The methods applied in the data normalisation step, which essentially is one that creates estimates of an 'alternative reality', are the same estimating methods outlined in section 2.2 comprising analogous, parametric and 'trusted source' adjustments.

1.4 Prepare the estimating plan

1.4.1 Agreeing the estimating approach(es) to be used

Using the information collated and the scope definition, the estimator can develop the estimating approach for each item. The approach selected will be influenced by the type of information, the number of data points, its maturity and applicability to the estimate. To ensure acceptance of the overall estimate, the approaches proposed should be agreed with key stakeholders. It is rarely necessary, or efficient, to agree the approach to each individual constituent estimating element; a certain level of estimator autonomy is desirable. However, at a holistic level, agreement on the approaches to be used will facilitate stakeholder acceptance of and confidence in the outputs.

1.4.2 Agreeing the competency requirements of the team

Having defined the scope of work and the estimating approaches to be used, the team structure, competence and availability can be assessed. The required skill sets for the estimating activities can be aligned to individual team members. It is common for highly experienced team members to have multiple responsibilities, so a sub-team may be constructed to support the estimating activities and dependencies while the team's expertise is utilised in the most effective manner.

Depending on the size of the company and the complexity of the project, the team structure may vary from a sole project manager, to a distributed team across the various functions. For large, complex projects it is recommended that the project has an impartial, qualified, senior estimator to oversee the estimating activities.

1.4.3 Agreeing the schedule of estimating activities and project dependencies

The scheduling of the agreed estimating activities must take into consideration the volume of work, and the size and capability of the team. In addition, when scheduling the estimating activities, it is key to understand any dependencies for those activities to take place, as well as dependencies on the estimating outputs, for example business case timelines. The timescales required for assurance and stakeholder briefings are frequently underestimated, so sufficient time must be allocated for these activities.

1.4.4 Updating the estimating RAM

Now that all the data has been collected, activities planned and resources allocated, the estimating responsibility assignment matrix (RAM) can be updated with the resources that will be employed, along with their areas of responsibility. This is a live document that develops as the estimating activities progress.

1.4.5 Defining the toolsets and information communication media

If toolsets are not available in time to support the estimating schedule, the outputs could be at risk. Defining these before the estimating activities begin will help mitigate any issues and ensure that the team is supported by the right tools.

A definition of the communication media for information is key to ensure that the team understands how this will be undertaken.

2

Creating the base estimate

2.1 Estimate management

2.1.1 Iterative process

Before starting any cost estimate, it is important to understand that it will be an iterative process. It is extremely unlikely that all information will be available at the same level of maturity at the start of the project. This means that the estimator must manage the estimating process as an iterative one. The estimator will progress through the various stages of the estimating process (see Figure 0.1: Estimating framework), and at each stage the individual's knowledge will develop as more up-to-date information becomes available. If the estimator plans for an iterative process and operates configuration control, it will facilitate these changes in a controlled manner.

2.1.2 Configuration control

A key principle of cost estimating is the ability to trace and document all aspects of the analysis from raw data to final outputs. To achieve this, robust configuration control is required. Considering that estimating is an iterative process, configuration control will provide a method to ensure that the most current information is used. It is important to tailor the extent of configuration management required. For a simple project with a small number of data points, a simple naming convention could be sufficient to control the configuration (e.g. yyyy.mm.dd [name] v1.n). Whereas for more complex projects, full baseline management and change control may be required to ensure all information is configured appropriately. APM provides comprehensive guidance on configuration management which can be tailored to the needs of the estimating process; see *APM Body of Knowledge 7th edition* (APM, 2019).

2.2 Estimating approach

An estimating approach is the direction, or means of arriving at an estimate, and to some degree implies the level of detail at which the estimate is created. With complex projects, it is often considered to be good practice to create an estimate using more than one approach as a means of providing a greater level of confidence in the output advised, thereby testing the robustness and interpretation of the data, the assumptions and the methodologies employed.

2.2.1 Top-down approach

In a top-down approach to estimating, the estimator reviews the overall scope of a project in order to identify the major elements of work and characteristics (drivers) that could be estimated separately from other elements. Typically, the estimator might consider a natural flow down through the work breakdown structure (WBS), product breakdown structure (PBS) or service breakdown structure (SBS).

The estimate scope may be considered as a whole, or broken down to different high levels of WBS as required (see Figure 2.1). The overall project scope must be covered by the range of non-overlapping work packages selected, although not all work packages need to be estimated at the same level of WBS. This allows the estimator to use the maturity and/or uncertainty in the key information available to produce the most appropriate level of estimate. The overall project base estimate would be created by summing these high level estimates. This should not be confused with the bottom-up approach where all the lower levels would be aggregated.

Over the life of the estimate these higher-level work packages and the associated higher-level estimates may be broken down into more detailed or more refined elements, which ultimately will facilitate a bottom-up approach (see Figure 2.2).

A top-down approach is frequently used for creating rough order of magnitude (ROM) estimates, otherwise known as ball-park estimates, where the level of detail available is limited. As a general rule, a top-down estimate requires less time and effort to produce than one produced using a bottom-up approach. Top-down estimates are appropriate at the beginning of the life cycle when large numbers of alternative options need to be estimated and considered. As the solution matures and more information becomes available, there is an increased opportunity to produce bottom-up figures. However, a top-down approach

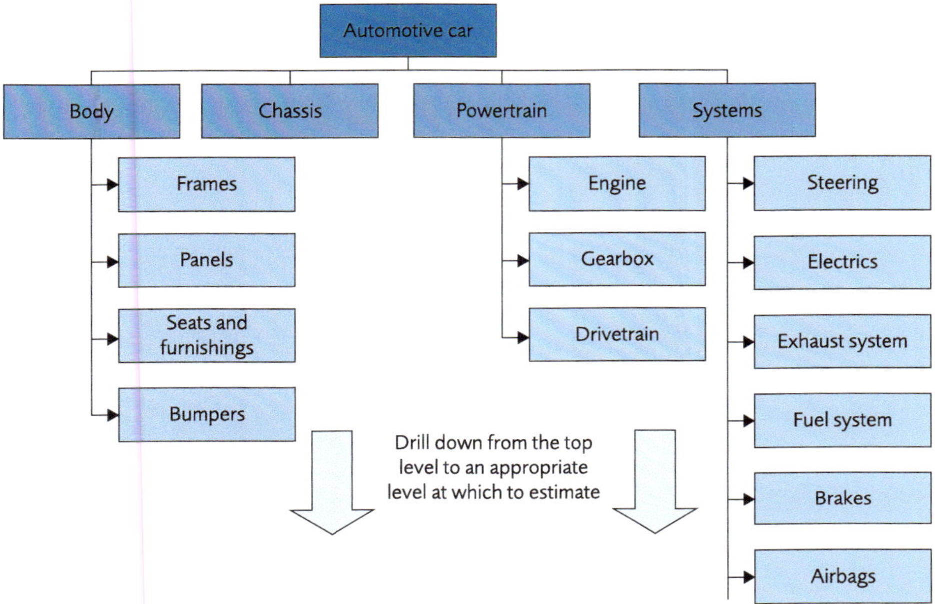

Figure 2.1 Top-down example
Source: © *Alan R Jones 2019*

can still be useful throughout the life cycle of a project, e.g. for validation purposes.

The main benefit of working at a higher level is that there is a tendency to use more holistic data from previous projects or products, including unmitigated and unforeseen risks, and scope creep. This can reduce the risk of emerging work activities or costs being overlooked. As a result, top-down estimates are typically greater than those created by a bottom-up approach.

Base estimates created by a top-down approach should exclude consideration of additional risks and opportunities. These should be considered separately by either a top-down or bottom-up approach as part of the formulation of the project baseline estimate. See section 3.

It is considered good practice to express an uncertainty range around a top-down estimate, based on the maturity of the information available, and the estimating methodologies employed. According to the *NAO survival guide to challenging costs in major projects*, (National Audit Office, 2018) "Early cost estimates should be presented as a range and never a point estimate". Note that APM and ACostE believe this should apply to all cost estimates.

Project life cycle				
Pre-concept	**Concept & assessment**	**Development/ demonstration**	**Manufacture & entry into service**	**In service operation & support**
Large number of different types of platform options	Large numbers of design options	Numbers of sub-systems options	Suppliers options	
Top-down			**Bottom-up**	

Figure 2.2 Project life cycle

2.2.2 Bottom-up approach

In a bottom-up approach to estimating, the project team interprets the client's requirement into a breakdown structure, identifying the lowest level at which it is appropriate to create a range of estimates covering the project scope based on the task definition available, or that can be inferred. A bottom-up approach requires a good definition of the task to be estimated, and is frequently referred to as detailed estimating, grass-roots or as an engineering build-up. Where the task definition has been derived, for example, through programme management and technical experience, rather than defined explicitly by the customer, the assumptions made should be recorded in the basis of estimate.

The level of detail available will be influenced by the maturity of the project, product or service and the level at which actual data, such as costs, is collected in the organisation. For very immature projects with little definition, a bottom-up approach may not be appropriate.

The overall project base estimate is compiled by aggregating the non-overlapping lower level estimates (see Figure 2.3).

Base estimates produced by a bottom-up approach are often favoured in some organisations because the levels at which the constituent estimates are compiled are more tangible than those created at a higher level through a top-down approach. Consequently, bottom-up estimates are often used to gain stakeholder confidence and buy-in. However, as the approach focuses on the aggregation of estimates of discrete packages, there is a higher risk that these will exclude any allowance for emergent work (scope creep).

In many organisations, bottom-up estimates may be based on the most likely values for each low-level activity, and as a consequence, it is not unusual for bottom-up estimates to be less in value than their equivalent top-down estimates. Typically, lower level baseline estimates are positively skewed and 'most likely values' are inherently and implicitly optimistically biased.

In simple terms, data is positively skewed if the most likely value (mode) is closer to the minimum or optimistic value than it is to the maximum (or pessimistic

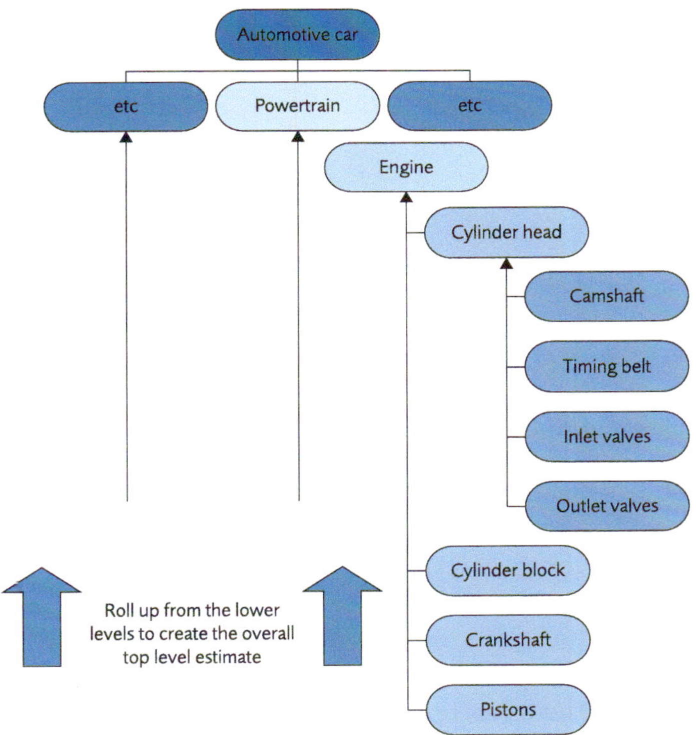

Figure 2.3 Bottom-up example
Source: © *Alan R Jones 2019*

value). In such circumstances the median (50 per cent confidence level) is greater than the most likely value, and the mean or average value is greater than the median. It is possible to cite the median or mean values as the central point of the 3-point estimate, but this should be clearly recorded in the basis of estimate so that it can be taken into account appropriately during the evaluation of risk and opportunity (section 3); this may guide the project manager to setting an appropriate project management reserve.

Due consideration to the 3-point estimate range should be made, based on the maturity of the information available, and the estimating methodologies employed.

Base estimates created by a bottom-up approach should exclude consideration of contingency, risks and opportunities. The base estimate should include the resources necessary to conduct the work described, adopting normal working practices and business norms. Risks and opportunities should be considered separately by either a top-down or bottom-up approach as part of the formulation of the project baseline estimate. See section 3.

2.2.3 Relying on the work of others – 'ethereal' approach

In some cases it is necessary to accept estimated values into the estimating process, the provenance of which is unknown and at best may be assumed. These are values often created by an external source for low-value elements of work, or by other organisations with acknowledged expertise. These could be through:

- vendor price quotation (either by tender or single source);
- catalogue prices;
- negotiated prices;
- subject matter experts.

Whether the external source has created the estimate by either a top-down, bottom-up, or even an ethereal approach (Jones, 2019 a), is immaterial if the basis of estimate is unknown. To the estimator or project receiving that input, it is un-auditable and is accepted on 'face-value'. Its provenance is unknown, having entered the system somewhat 'out of the blue'.

The ethereal approach should be considered the approach of last resort where low maturity is considered acceptable. The approach should ideally be reserved for low-value elements of work, and situations where a robust estimate is not considered critical. However, this approach may be the only one available in some special circumstances, as shown in Figure 2.4.

	Low	High
Unfamiliar product or service	For rough order of magnitude only where time to estimate is restricted or resource limited.	Single source specialist supplier (independently audited)
Familiar product or service	Commodity items where time to estimate is restricted or resource limited	For rough order of magnitude only where time to estimate is restricted or resource limited.

Relative value of product or service being estimated

Figure 2.4 Potential suitable use of ethereal approach

It is often the case that an estimate produced using an ethereal approach is input into the system as a single value. Some consideration of the uncertainty range around this estimate should be expressed by a 3-point estimate, taking account of the likelihood that the data may be positively skewed rather than symmetrical.

2.3 Estimating methods

An estimating method is a systematic means of creating an estimate, or an element of an estimate. There are only three basic estimating methods (as depicted in Figure 2.5):

- analogy;
- parametric;
- 'trusted source'.

Some sources may refer to other estimating methods such as 'extrapolation from actuals', but these are fundamentally specific cases of analogy, parametric and trusted source methods, or a combination of all three.

Engineering build-up is also often described as a method but it is really a specific bottom-up approach. Engineering build-up will typically draw on a combination of analogy, parametric and trusted source methods and a variety of detailed techniques.

Simulation could be considered as a fourth estimating method, but for the purposes of this guide it will be considered as a group of related techniques that

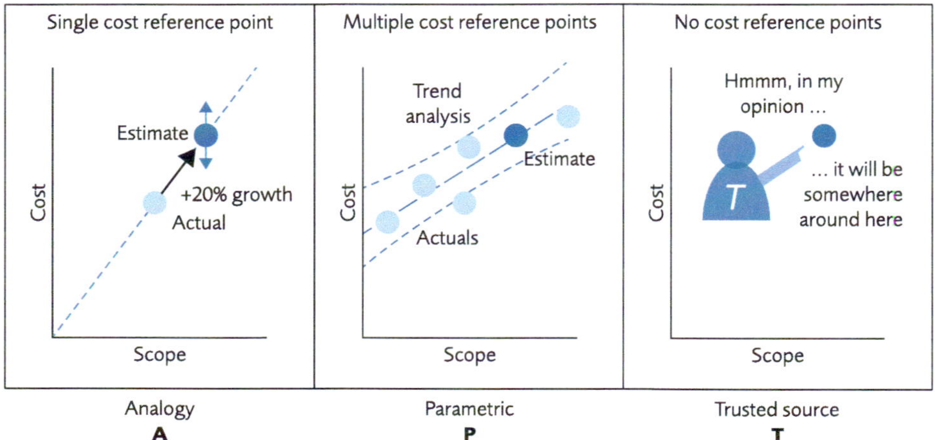

Figure 2.5 Estimating methods
Source: © *Alan R Jones 2019*

might be used to support (typically) a parametric method. However, a one-off simulation can be used as a technique to support an analogous method.

Regardless of the method adopted, there will be an inherent uncertainty in all base estimates. This is due to the implicit use of assumptions within the defined scope, and the lack of certainty around future performance, skills and competence of those discharging the emergent work content. It is considered good practice to develop a 3-point estimate for discrete elements of work, regardless of the estimating method selected. The 3-point estimate should consist of:

- **A minimum value**: A value that might realistically be achieved with higher than normal performance levels or lower than anticipated complexity in the emergent scope definition. This is not necessarily the absolute minimum value, but it may be considered appropriate to use the lowest recorded incidence of the same scope and complexity of work.
- **A most likely value:** A value that might realistically be achieved with normal observed or expected levels of performance, and the expected complexity in the emergent scope definition, assuming normal working practice.
- **A maximum value:** A value that might realistically be achieved with lower than normal performance levels or higher than anticipated complexity in the emergent scope definition. This is not necessarily the absolute maximum value, but it may be considered to be appropriate to use the highest recorded incidence of the same scope and complexity of work.

All assumptions made should be consistent with the authorised project assumptions and new, lower level assumptions should be recorded in the detailed basis of estimate (BoE).

2.3.1 Analogy

Estimating by analogy is a means of creating an estimate by comparing two similar entities (e g. projects, products, widgets, etc.). One entity is used as the reference point against the other, and rational adjustments are made for differences between the two. Analogous estimates are sometimes referred to as 'comparative estimates' or 'reference class' forecasts. The critical aspect of this methodology is that the analogous items selected must have similarities or characteristics to enable them to be estimated through the consideration of a number of key variables of a technical or programmatic nature.

It is important that we recognise that the very simplicity of any analogical estimating method implies an underlying estimating relationship or metric that passes through the origin.

Analogous techniques are generally limited to factoring and one-off simulation.

For example, for a two-factor analogy:

In the example in Figure 2.6, the relative cost contribution assumes that the weight is a primary driver, whereas the number of components is a secondary driver. The weighting may be industry or business specific.

Product	Assembly A	Assembly B	Cost driver change factor (B:A)	Relative cost driver weighted contribution	Change factor (B:A) weighted by cost contribution
Cost driver 1: weight	100 kg	80 kg	80%	75%	60%
Cost driver 2: no. of components	50	60	120%	25%	30%
Total or net cost driver contribution	100%	90%	←	←	90%
Actual cost/ estimated cost	£2,000	£1,800			

Figure 2.6 Example two-factor analogy

2.3.2 Parametric

A parametric estimating method is a systematic means of establishing and exploiting a pattern of behaviour between the variable that we want to estimate, and some other independent variable or set of variables or characteristics that have an influence on its value.

The fundamental difference between parametric estimating and analogous estimating is that parametric estimating is based on an assessment of more than one data point. This data is either collated at the time the estimate is being created or based on a general form of relationship that has been previously determined, e.g. a standard or norm. As the estimate will be based on a number of past actual observations, it will be possible to draw some statistical inferences on the robustness (or otherwise) of the estimate produced.

The majority of detailed numerical techniques fall within the parametric methodology.

Examples include the use of standards and norms, learning curves, Norden-Rayleigh curves (Turré, 2006), Chilton's Law (Norden, 1963), regression analysis, time series analysis.

For example, suppose two potential cost drivers (weight and number of tests) have been identified for a particular product based on previous similar products. The evidence depicted by plotting the cost of each product by the potential cost driver value as shown in the upper right and lower left graphs of Figure 2.7 support this. By plotting the historical data for these cost drivers against each other, the degree of scatter in the upper left graph of Figure 2.7 suggests that there is only a weak relationship or correlation between them (i.e. they are not duplicating each other's contribution to cost.) The lower right-hand graph of Figure 2.7 shows that a better fit to the data can be achieved using both cost drivers, based on a multi-linear parametric estimating technique rather than a simple linear parametric technique using a single cost driver. The arrows on Figure 2.7 show the flow of logic.

2.3.3 Trusted source

A 'trusted source' method (Jones, 2019 a) is a means of capturing values for which we either have no obvious historical reference point, or for which the expertise is vested in others and the estimator has little control or influence in being able to validate or challenge the basis of estimate.

The most obvious example of a trusted source method is that of 'expert judgement', but it might also include the use of vendor quotations and commodity

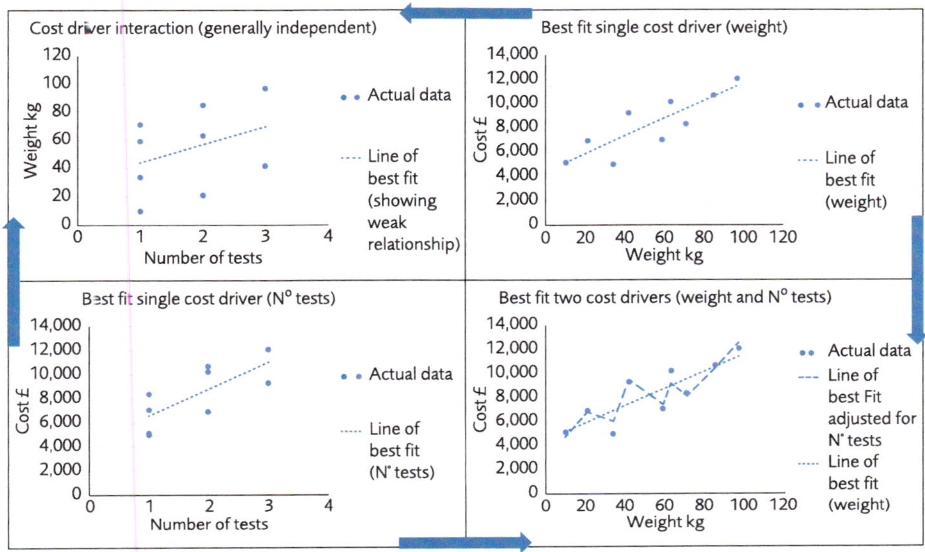

Figure 2.7 Example parametric with two cost drivers
Source: © *Alan R Jones 2019*

prices, or a commercial estimating toolset. The trusted source method is often synonymous with an ethereal approach (see section 2.2.3).

2.4 Documenting the basis of estimate (BoE)

A basis of estimate (BoE) is a record of the information and logic used to underpin an estimate. The BoE document should provide enough relevant details to fully understand 1) the scope of the project; 2) the information used to compile the estimate.

The BoE provides an auditable record of all the data used to deliver the estimate to a defined scope; therefore it should be able to track the changes made to the estimate throughout the lifetime of a project. It is important that the key data sources, assumptions, exclusions and dependencies are included.

The type and precision of the estimate should be included; this is likely to change as a project gains maturity through its life cycle as more information becomes available, with the consequence of the accuracy levels increasing.

The estimate plan should be outlined within the BoE with dates of when the estimate is planned to be completed. The BoE will include the product (or service) breakdown structure (PBS), work breakdown structure (WBS) and cost breakdown structure (CBS) for the project to determine where the major costs are located. There should be traceability between the PBS, WBS and CBS.

Due to the inherent uncertainty in any estimate value and the information used to compile it, recommended practice is to express the uncertainty through three points (minimum, most likely and maximum).

The general structure of a BoE is shown below. Its contents will depend on the size and maturity of the project, but it should generally include:

- project scope;
- estimate plan;
- project execution responsibilities and strategy – who delivers what;
- product breakdown structure;
- work breakdown structure;
- scope information to support required estimate precision / accuracy;
- approach to compiling the estimate;
- use of norms and benchmarking;
- cost breakdown structure;
- estimating uncertainty ranges;
- assumptions, clarifications, qualifications and allowances;
- references;
- review and approval signatures.

See glossary for definitions: precision, accuracy and uncertainty.

During contract price negotiation or any type of internal or external review, it is considered to be essential practice that there is a formal record of any change that may arise as a result of such negotiation or review. These changes may refer to the estimated cost, the scope of work, or any underlying change in assumptions, dependencies, opportunities, risks and exclusions (ADORE) and should be recorded in the basis of estimate.

Any estimating support for contractual change should reflect the good practice outlined in the rest of this guide. The size and nature of the change may influence the choice of approach or method, and the results recorded in the estimate configuration control log.

3

Risk, opportunity and uncertainty assessment

3.1 Assessing uncertainty in the baseline activities

There will always be uncertainty (or a lack of exactness) in the cost estimate of projects. This happens for various reasons, including uncertainty about the detailed scope of work and uncertainty about the levels of productivity that will be achieved. In addition to these factors, there will be risks and opportunities; these are things that may or may not happen, but if they do they will impact the estimate. Risks are discrete events that will have a degree of uncertainty over the exact value. Baseline tasks will occur, but the actual value will have uncertainty. For example, while driving home (baseline task), the journey time is ***uncertain*** due to variable traffic conditions (i.e. traffic lights). However, in addition there could be a ***risk*** of a delay of ***uncertain*** duration due to an accident.

Uncertainty is the inherent and potentially uncontrollable variability in estimating the actual cost and schedule. It can be considered as a tolerance band on the understanding of the scope. Uncertainties arise because the organisation does not have a complete understanding of the proposed task or the solution. An uncertainty is an expression of something that will happen; the actual project value will not be known but is expected to lie within a defined range. Some uncertainties will express natural variation; for example, my journey home each day varies by maybe five minutes less or 10 minutes more.

Most baseline tasks in a work breakdown structure will have uncertainty, which can be expressed as three values: the minimum (unlikely to be less than), most likely value and maximum (unlikely to exceed). The 3-point estimate should express the range of uncertainty – excluding risks or opportunities.

Where there is uncertainty in the scope definition of the baseline activities, the 3-point estimate can be used to express the extremes of a minimum (or simplistic scope requirement) and a maximum (or complex scope requirement). Alternatively, any potential but improbable extreme in the scope requirements might be expressed as a risk or opportunity. Care must be taken not to duplicate or overlap any extremes expressed in both ways.

3.2 Link to the risk management process

The evaluation of the net effect of risks and opportunities cannot be performed in isolation from the baseline activities, nor can they be evaluated in isolation from the project's risk management process (i.e. do not re-invent the wheel). Most of the data requirements to manage risks and opportunities are also needed to evaluate their net impact; for example, 3-point estimate of the individual risk or opportunity cost impact, probability of occurrence, risk retirement date, mitigation plan, etc. In addition to these parameters, in order to evaluate their net impact, there is a need to express the 3-point estimate range of potential values as a probability distribution. This then allows probabilistic modelling of the risks and opportunities to be performed in Monte Carlo simulation in conjunction with baseline activities.

Risks and opportunities are discrete events whose occurrences are expressed by a probability of occurrence, modelled by a Bernoulli distribution.

The most common forms of probability distributions to model uncertainty include normal, lognormal, exponential, triangular and uniform.

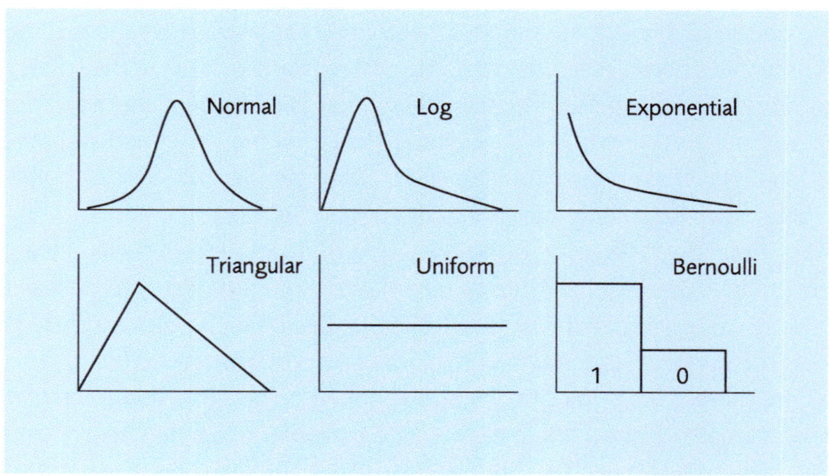

Figure 3.1 Different types of distribution

When to use these different types of distribution:
Normal distribution:

This symmetrical distribution often represents the spread and frequency of values in naturally occurring observations in nature, such as the height of adult males or females of a given ethnicity. It can also represent the distribution of values from man-made systems such as the accuracy and/or precision of machining operations. The distribution can often be used to represent the range of values for system or sub-system level costs, even where the constituent elements of those systems or sub-systems are not normally distributed. The scatter or deviation of values around a linear cost estimating relationship is also expected to be normally distributed.

Lognormal distribution:

This distribution can be used in reliability analysis to model the repair time of items, in particular in relation to the fatigue-stress characteristics of mechanical systems. It is often an empirical distribution observed in natural growth or human behaviour systems. When a variable is deemed to be lognormally distributed in linear space, its values will be normally distributed in logarithmic space. As a consequence, the scatter or deviation of values around a power or exponential cost estimating relationship is expected to be lognormally distributed.

Triangular distribution:

The triangular distribution is frequently used as a default distribution where there is some knowledge or perception of a most likely value, and also an appreciation of the likely minimum or maximum values of a variable. In cost and schedule scenarios, data is more likely to be positively skewed; i.e. where the difference from the most likely to the maximum is greater than the difference between the most likely and the minimum. Aggregated system level variables are more likely to be symmetrically distributed.

Uniform distribution:

The uniform distribution is frequently used as a default distribution where there is some knowledge or perception of the likely minimum or maximum values of a variable, but no evidence or knowledge of the most likely value, i.e. any specific value being more likely than any other. For example, random numbers are uniformly distributed.

Exponential distribution:

The exponential distribution is used to represent the inter-arrival times between random events in a queueing system, e.g. the actual arrival time between buses, demand for spares or the need for repairs. It can be used in reliability analysis where there is a constant failure rate. It is referred to as a 'memoryless distribution', as the time to the next event occurring is independent of the time since the last event.

Bernoulli distribution

The Bernoulli distribution is used to simulate the probability of occurrence of individual risks and opportunities in Monte Carlo simulation. It functions as a binary on/off switch such that the proportion of the time that the distribution returns a value of one (the 'on' condition), matches the defined probability of occurrence of the risk or opportunity in question.

When selecting a probability distribution, it is worth considering the following points.

- In practical terms, a subject matter expert may be able to advise on the input parameters for various distributions (minimum, most likely and maximum). It is recommended, where possible, to use historical data to determine the ideal distribution.
- The extra precision you get from using the more specific distributions, for example lognormal versus triangular, may not result in a materially significant gain.

3.3 Link to schedule risk analysis

Schedule risk analysis is similar to cost risk analysis, in that the estimator will need to allocate risks, opportunities and uncertainties to activities within their schedule. If the cost estimate is derived from the schedule estimate, there will be a link between schedule duration and cost (see Figure 3.2).

A project that is given more schedule than needed is likely to be more expensive, as tasks expand to fill the time available. However, inappropriate schedule compression can have a significant impact on the project cost. A compressed programme will require more management effort, more resource movement, a high level of turbulence and a propagation of risk through the project as tasks may be forced to run concurrently.

See also 3.4.1 Top-down approach, below.

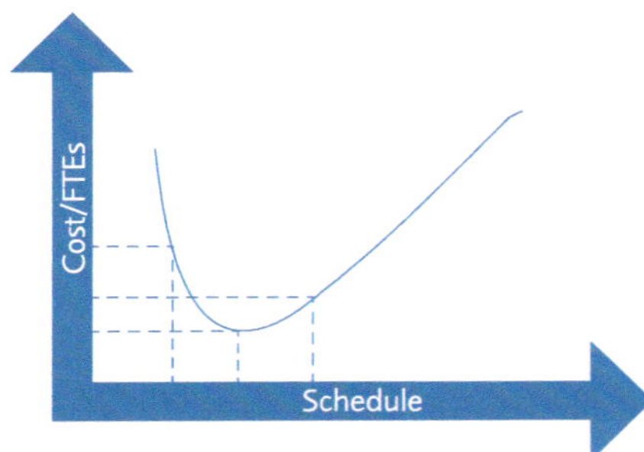

Figure 3.2 Schedule risk analysis

3.4 Holistic review of risk, opportunity and uncertainty

There will always be uncertainty (or a lack of exactness) in the cost of projects for various reasons, including uncertainty in the detailed scope of work and uncertainty in the levels of performance that will be achieved. In addition to these, there will be risks and opportunities, which are things that may or may not happen, but if they do their impacts will also bring a level of uncertainty.

A holistic approach to the potential impact of risks, opportunities and baseline uncertainties can be gained by comparing a top-down approach with a bottom-up approach to the evaluation.

3.4.1 Top-down approach

A top-down approach can be derived in several ways. If the organisation maintains good configuration control of its estimates, forecasts and outturns, one option would be to assess the typical uplift factor (or risk metric) between initial baseline 'most likely' estimates and confirmed outturn positions for different types of project. However, this may be better used as a check value to validate a value produced by an alternative approach and method.

For a more pessimistic approach, cost overruns might be assumed to be proportional to schedule slippage on the pretext of 'time is money' (Labaree, 1961), and that resource that has been deployed onto a project can be difficult to reassign elsewhere temporarily. On this basis, for example, a 25 per cent slippage in schedule could result in a 25 per cent increase in cost (see Figure 3.3). This is sometimes referred to as the Marching Army Technique (Jones, 2019 c). In conjunction with this, there will be some costs that are not correlated with schedule, such as some material costs, escalation or foreign exchange; this will allow a pessimistic scenario to be developed by factoring these costs.

It is recommended that any pessimistic, top-down approach of this type (or 'glass half empty perspective') is compared with the optimistic, bottom-up approach often produced through Monte Carlo simulation. However, if a robust

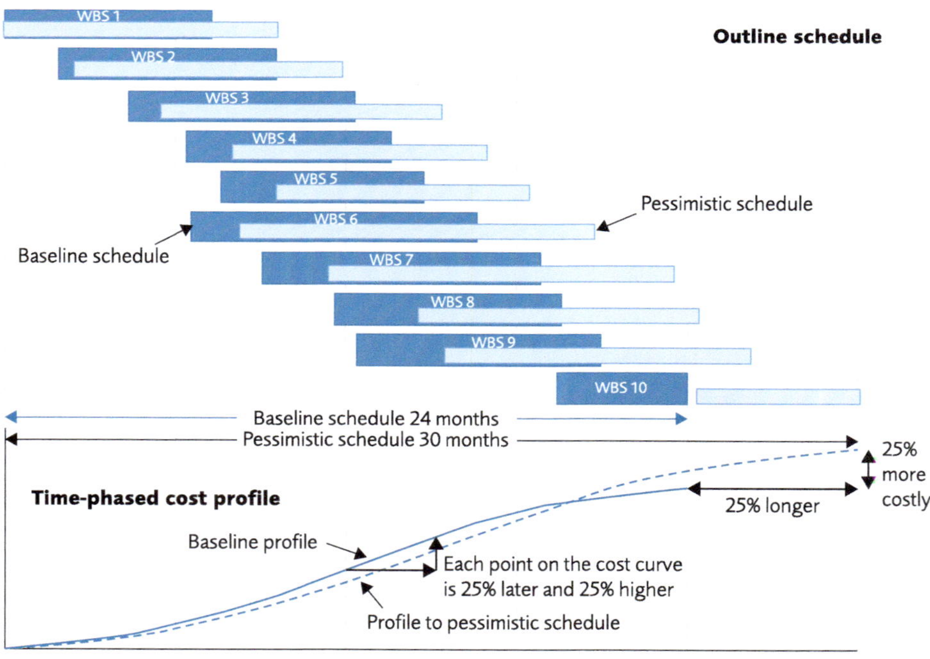

Figure 3.3 Time phased cost profile
Source: © *Alan R Jones 2019*

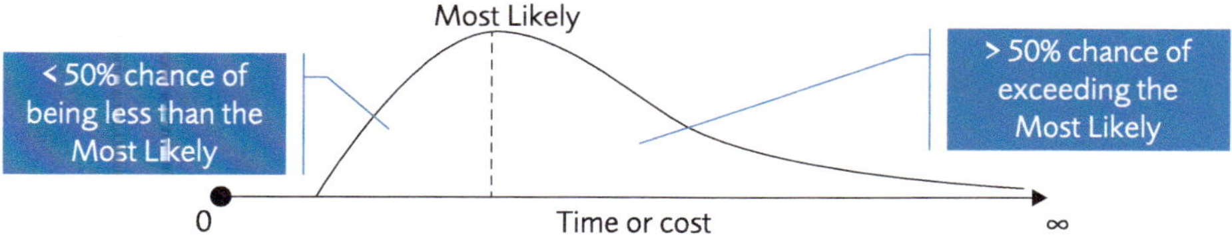

Figure 3.4 Realistic distribution

risk and opportunity management plan or register is not available, then any pessimistic, top-down approach should be compared with a value derived by a typical risk metric uplift factor.

Note that there is an inherent expectation that the negative impact of risks will be greater than the positive impact of opportunities, as the former is unbounded to the right and the latter is bounded to the left (i.e. we cannot have negative costs or schedule) (see Figure 3.4).

3.4.2 Bottom-up approach – Monte Carlo simulation

Where there is a detailed risk and opportunity management plan or register that has been cleansed to remove duplications and baseline task uncertainties, a bottom-up evaluation of risk, opportunity and uncertainty can be created. Often the technique used will be Monte Carlo simulation.

Monte Carlo simulation performs analysis by building models of possible results by substituting a range of values, a probability distribution, for any aspect of your estimate. It then calculates results multiple times, each time using a different set of random values from the probability distributions.

During each Monte Carlo simulation, values are sampled at random from the probability distributions. Each set of samples is called an iteration, and the resulting outcome from that sample is recorded.

Monte Carlo simulation does this thousands of times, and the result is a distribution of possible outcomes. In this way, Monte Carlo simulation provides a much more comprehensive view of what may happen. Monte Carlo tells you not only what could happen, but how likely it is to happen.

Figure 3.5 above shows a typical outcome from the simulation. The confidence (percentile) is shown as a line in the range 0–100 per cent. It is the probability of achieving a budget, i.e. low confidence of achieving a low budget (left on this diagram) and high confidence of achieving (being better than) a high budget (right on this diagram). The 50 per cent confidence level would lie in the middle

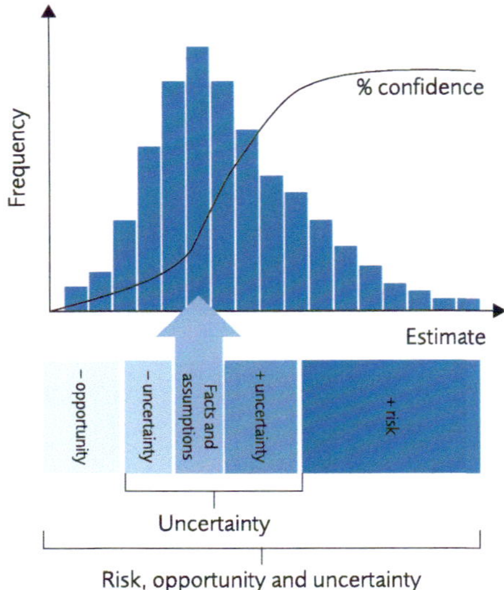

Figure 3.5 Monte Carlo

of the estimate and would represent the median value. 'Confidence level' is sometimes referred to as 'percentile'. The P50 represents the 50 per cent percentile or the 50 per cent confidence level.

It is imperative that risks, opportunities and baseline uncertainties are modelled as a single, interactive system in a Monte Carlo simulation analysis, before any conclusion is drawn. However, it is acceptable to run the baseline uncertainty models or risk and opportunity models independently in order to identify and manage the key drivers in the overall system.

Refinements can be made to the Monte Carlo model by mitigating risks and enhancing opportunities, and including these activities in the baseline. It is important to understand that a mitigation plan or action may not fully eliminate the risks, resulting in a residual risk exposure.

Care should be taken to avoid assuming that all tasks are independent of each other; doing so would result in excessive narrowing of the Monte Carlo model results. Therefore, due consideration should be given to a level of background correlation between baseline tasks to avoid any such excessive narrowing. This narrowing is a natural reflection of the empirical observation that not all the good things in life happen at the same time, and nor do all the bad things. To counter this, the application of correlation allows for relationships between tasks to move together in the same direction without being intrinsically linked.

Maturity of task definition and performance	Likelihood of occurrence: Baseline tasks which **will occur**	Likelihood of occurrence: Risks and opportunities which **may or may not occur**
Unknown or poorly defined	**Unknown knowns** We know we have to do the task but its exact scope is not clear, ... or we know the task but we do not know how well we will perform it	**Unknown unknowns** These are those genuine risks and opportunities that we have not considered because they haven't occurred to us
Known or well defined	**Known knowns** We know we have to do the task and we are clear of the requirements and understand our likely performance.	**Known unknowns** We have identified tasks that may or may not need to be carried out. These are our defined risks and opportunities.

Figure 3.6 Risk, opportunity and uncertainty evaluation
Source: © *Alan R Jones 2019*

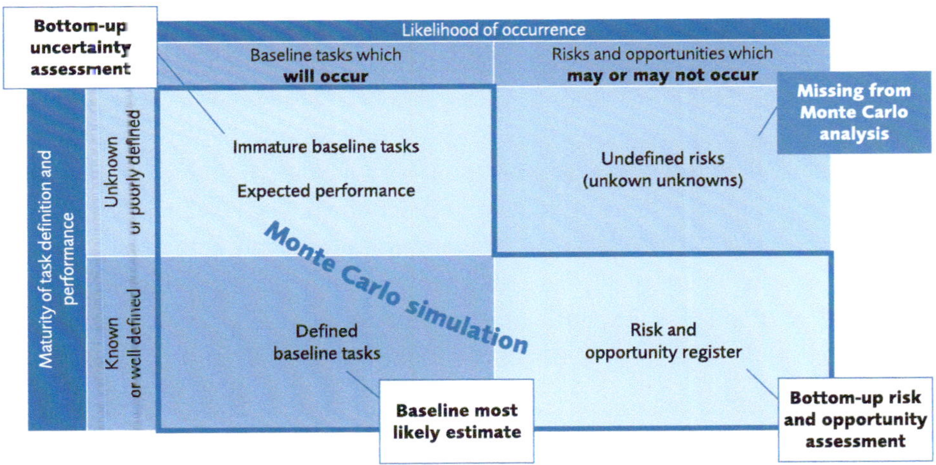

Figure 3.7 Monte Carlo optimistically biased
Source: © *Alan R Jones 2019*

It should be noted that the outturn from a Monte Carlo simulation is inherently optimistically biased, as it implicitly excludes any unknown risks, referred to as unknown unknowns'.

From Figure 3.6 (Jones, 2019 c), the unknown unknowns are not included in a Monte Carlo simulation, as in Figure 3.7.

When not using Monte Carlo, a risk factoring technique is sometimes employed in which the most likely risk value is factored by the probability of occurrence. This results in an optimistic value on two counts:

- In general terms, the sum of the factored values most likely will understate the true modelled mean value.
- It excludes any 'unknown unknowns'.

For low probability, high impact risks, these may have to be revisited and potentially moved to 'Exclusions'. It is not appropriate to create a risk provision for such risks by a factoring technique.

For example, if there is a low probability of a lightning strike, which results in total catastrophic failure, you cannot factor for a percentage based on probability and impact; you will need to allow for a total replacement. Hence it would be excluded from the estimate or be included at the full value.

3.4.3 Dealing with inherent optimism bias in risk and opportunity

As outlined in Figure 3.7, Monte Carlo simulation is inherently optimistically biased; a 50 per cent confidence level (often referred to as P50) is not the true median. Therefore this guide recommends that a higher confidence level is taken for any single point estimate for cost and schedule, e.g. P80 might be more appropriate unless otherwise contractually required. The actual value would depend on the maturity of the risk environment e.g. register and baseline estimate, and the risk appetite of the organisation. Even in a mature risk environment, the P50 for a 50 per cent confidence level is not recommended; a higher value is advised.

It is recommended that the output profile of the risks, opportunities and uncertainties is used to ensure that the impact of significant risks is adequately covered. This can be achieved by comparing the optimistic, bottom-up approach with a pessimistic, top-down approach (Jones, 2019 c).

The Treasury's guidance on appraisal and evaluation (HM Treasury, 2018) is mandatory for government projects. It states that there is a demonstrated, systematic tendency for project appraisers to be overly optimistic. It recommends that adjustments are made for optimism, based on data from past projects or similar projects elsewhere, and adjusted for the unique characteristics of the project in hand. It also contains supplementary guidance valuing infrastructure spend. This supplementary guidance gives advice for the realistic estimation of contingency budgets in programmes with mature cost estimating processes, while the main HM Treasury's *Green Book* sets out a simpler approach to optimism bias for less mature organisations.

3.4.4 Taking a balanced view of risk, opportunity and uncertainty

A top-down approach to risk, opportunity and uncertainty often generates a pessimistic perspective of the overall project estimate, but a bottom-up approach using Monte Carlo simulation is inherently optimistically biased. Often the project life cycle will determine the approach used; for example, in the early stages, a top-down approach may be more suitable when there is little information available and the uncertainty is high. However, it should be recognised that a top-down approach may also be used throughout the life cycle to validate the bottom-up approach. This allows a balanced view to be taken of the realistic level of risk, opportunity and uncertainty. This probably falls somewhere between the two extremes.

From a pragmatic perspective, a balanced view may be achieved by reviewing the nominal confidence level on the Monte Carlo simulation output equivalent to the top-down value calculated, and considering a realistic confidence level to be

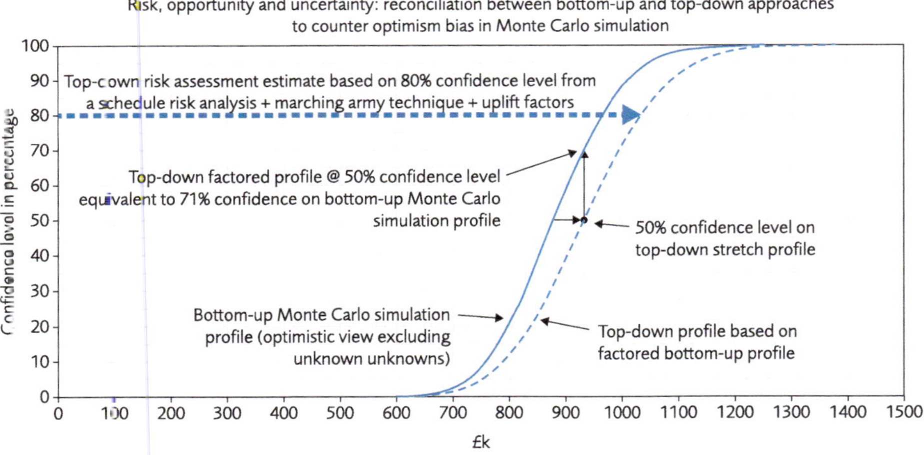

Figure 3.8 Comparison of bottom-up and top-down approaches to risk, opportunity and uncertainty
Source: © *Alan R Jones 2019*

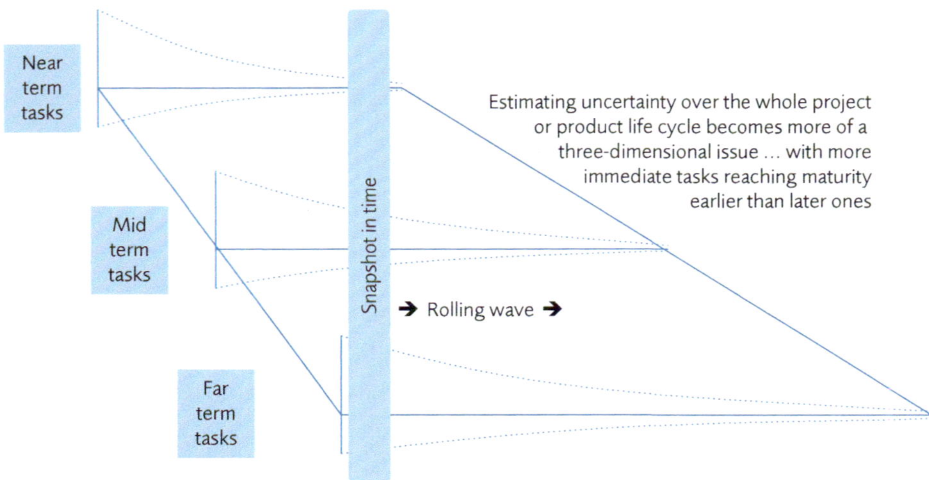

Figure 3.9 Rolling wave estimating
Source: © *Alan R Jones 2019*

in the range of greater than 50 per cent and less than the nominal top-down confidence level. The value chosen should ideally take account of the profile of the known risks and opportunities to ensure that the cumulative risk is adequately covered by an acceptable level of contingency above the agreed base estimate value (see Figure 3.8).

Rolling wave estimating is a technique in which estimators develop more precise, short-range estimates and less precise, longer-range estimates (see Figure 3.9). The near term tends to have less risk and uncertainly than the longer term. Rolling wave estimating is a continuous process of developing and managing the estimate over time, and it is the epitome of an iterative process.

4

Overall estimate validation, challenge and approval

If the organisation is going to place reliance on an estimate, it needs to have quality assurance arrangements in place. This must include an external review of the data and methods used with the resulting outputs. This guide has emphasised that estimates at different stages of a project will have differing levels of maturity. The first part of this section sets out how to define and assess estimate maturity. It then covers the need to review estimates prior to approval.

It is good practice that all estimates are subjected to an appropriate level of validation or challenge prior to approval. This should consider the assumptions, dependencies and exclusions captured in the ADORE or MDAL statement in relation to the creation of the base estimate, as well as the impact of risks and opportunities, to ensure that there is consistency in their interpretation, and that any values input and output are both credible and supportable. The validation and challenge of the base estimate may be performed before the evaluation of risks and opportunities, as the review of the former may result in new risks or opportunities being identified or existing ones removed.

It is important that procedures are carried out to check the data inputs, assumptions, methods and outputs of the cost model used. It is equally important to capture lessons (see Figure 0.1) and feed them into future estimates. To provide a consistent level of scrutiny there needs to be an impartial body conducting these reviews.

4.1 Estimate maturity review

4.1.1 Maturity definition

Estimate maturity levels are a convenient way to define the suitability of an estimate for use. As described in section 1.2.2, an estimate's maturity can be expressed in a range, e.g. one to nine, where a maturity level of one would represent a low maturity estimate and nine would represent a high maturity estimate.

The concept of estimate maturity is based on the following principles.

- The estimate must be fit for purpose: Not all estimates need to be developed to the same maturity – the maturity can be a reflection of the intended use (purpose) of the estimate. The same estimate can have a varying maturity over the life cycle of a project. For example, a business may be exploring concepts and want a quick, low-maturity estimate on which to decide the best option to take. However, if the business is committing itself to an external customer, the business would often need a high maturity estimate.
- The business can tolerate uncertainty: This means that the business can accept a lower maturity estimate under some conditions. A lower maturity estimate will, in general, have a higher level of uncertainty, i.e. we would expect a wider +/- precision. Conversely, a higher maturity estimate would have a lower uncertainty. It is not always appropriate for a business to accept only high maturity estimates.
- The process can be tailored: The level of effort needed to develop an estimate will tend to be less for lower maturity estimates than for higher maturity estimates. This means that the estimation effort should be no more and no less than required to generate an estimate of the required maturity. The estimation process will need to have mechanisms by which the estimator can 'tailor' their practices depending on the estimate maturity e.g. for a low maturity estimate, the estimator may only need to consider the major risks, but for a higher maturity estimate, the estimator will need to perform robust risk analysis.

Creating a concept of maturity levels means a business defines a simple way to express the desired maturity of an estimate in a language that can be easily communicated and recognised. Figure 4.1 is an example of how a business might

Maturity level	Estimate uncertainty (precision)	Description
9	−5%/+5%	Product realisation/critical design review Production readiness review
8	−9%/+10%	Contract signature/memorandum of understanding Full concept definition Preliminary design review/programme commitment review
7	−17%/+20%	Request for quotation Preliminary concept definition Business plan approval Concept review/business concept review
6	−29%/+40%	Request for proposal Innovation and opportunity selection Prospect readiness review
5	−44%/+80%	Request for information Innovation and opportunity selection Prospect pursuit review
4	−62%/+160%	Exploring trade options
3	−76%/+320%	Exploring strategy
2	−86%/+640%	Exploring expectations
1	−93%/+1280%	Do not use

Figure 4.1 Example of maturity definition

wish to define estimate maturity. This is only an example and a business would need to define and agree its own definition for maturity.

In this example, maturity levels of four and below are only used for exploratory studies and level five or above follow the project life cycle.

For wide ranges it is important that stakeholders understand that the range is due to the maturity and is not due to contingency, which must be dealt with separately; see section 3.4.4.

4.1.2 Maturity assessment

When the maturity levels have been defined, an organisation can then develop an objective scoring mechanism to assess the maturity level for a specific estimate, or sub-elements of it. There are many ways to 'quantify' maturity. Three methods are described, with an increasing level of objectivity.

Method 1: Simple checklist

At its simplest level, the scoring mechanism could be based on checklists (see Figure 4.2).

Level	Estimate based on ...
EMA9	Precise definition with recorded costs of the exact same nature to the estimate required
EMA8	Precise definition with recorded costs for a well-defined similar task to the estimate required
EMA7	Precise definition with validated metrics for a similar task to the estimate required
EMA6	Good definition with metrics for a defined task similar to the estimate required
EMA5	Good definition with historical information comparison for a defined task similar to the estimate required
EMA4	Defined scope with good historical information comparison to the estimate required
EMA3	Defined scope with poor historical data comparison to the estimate required
EMA2	Poorly defined scope with poor historical data comparison to the estimate required
EMA1	Poorly defined scope with no historical data comparison to the estimate required

Figure 4.2 Example of estimate maturity assessment

Source: BAE Systems

Question	Weight W	Score S	Weighted Score W*S	Description of maximum score (100%)
Is the purpose of the estimate understood?	0.5	60%	0.3	The purpose of the estimate is understood. The required maturity is defined and has been agreed with the customer and all the key stakeholders.
Is the project being estimated understood?	1.5	50%	0.75	The project requirements, scope and environment are understood. Concepts are agreed and requirements approved and baselined.
Is the estimation team competent?	0.7	100%	0.7	An estimating team has been formed and they are , suitable experienced and qualified. The team has received appropriate training and has the appropriate domain experience.
Have the estimating team roles and responsibilities been defined, agreed and committed?	0.5	50%	0.25	There is evidence that the estimating team members are aware of their roles, responsibilities and accountabilities. The team includes authors, reviewers and approvers.
Is the estimate based on credible historical data and/or calibrated tools?	1	25%	0.25	Any tools used and data sources have been documented. Both data and tools are suitable for the project being estimated.
Were sufficient time, effort and appropriate resources made available to generate an estimate?	0.6	50%	0.3	For the required estimate maturity and size of project the resources, time and effort used to generate the estimate is appropriate.
Was the estimate developed using robust estimating techniques?	0.4	50%	0.2	The estimation used recognised techniques and documented the logic and rationale for the approach taken.
Were project targets met and are the risks and compromises accepted?	0.5	50%	0.25	There is evidence that, where project targets are set, the estimate author found a low risk solution to meet those targets. Risks and compromises are understood and have been documented. The customer and budget approver find the level of risk and requirement compromise acceptable.
Was risk and uncertainty analysis performed?	1.8	50%	0.9	The risks, opportunities and uncertainties are understoodin detail by the estimator and are presented using a 3-point estimate. Statistical methods were used (e.g. Monte Carlo or equivalent) to generate a confidence level profile from which a 3-point estimate was established.
Was the estimate documented?	0.6	100%	0.6	The estimate author documented the estimate in accordance with the required format and content.
What was the maturity of the estimate review?	0.9	100%	0.9	A formal review was conducted by independent, competent people, was robust, had a terms of reference, was well attended, etc.
Total	9		5.4	

Figure 4.3 Example of maturity assessment

Question	Definition	0%	25%	50%	75%	100%
The project being estimated was understood	The estimator has an appropriate level of understanding of the project being estimated.	No definition.	Rough scheme. Outline design proposals.	Outline scope, specification and preliminary designs.	Scope and specification drafted. Outline design complete.	Proposal fully defined. Design complete, scope fully established, specification fully met in design.

Figure 4.4 Example of a maturity scale

Method 2: Weighted checklists

More advanced versions may use weighted checklists as shown in Figure 4.3. In this example, each question has been given a weighting and a level of compliance in meeting the requirements from 0–100 per cent. Zero per cent means the estimate scored no points, and 100 per cent means it scored maximum points.

Example of weighted scores. Note that the values given are for illustration purposes only.

The estimator scores their compliance to each question. Each question has been weighted by the business for its importance. The weighted score is the estimator's actual estimate maturity.

The total weighted score is then used to determine the maturity level in Figure 4.4.

Method 3: Score card

The weighted checklist method (method 2) relies on the assessor to estimate the level of compliance (or achievement) of each of the checklist questions. Would the same assessor score in the same way on two separate occasions? Would two separate assessors come up with the same scores?

We can introduce a higher level of objectivity by providing pre-defined answers to each question. The example below illustrates what a checklist question might look like for the example question, "The project being estimated was understood". Once an assessor has selected the best answer, the per cent compliance (achievement) is then automatically derived.

This type of scoring mechanism ensures the maturity assessment is more objective, repeatable and defendable. An organisation would need to define a range of answers for each maturity assessment question.

4.1.3 Why is maturity assessment important?

The use of maturity levels and maturity assessment ensures that all estimating maturity scores are objective, repeatable and quantifiable. Scenarios in which an estimating maturity assessment can be used include the following.

- The business needs a high maturity estimate; the assessment will evaluate if the estimate is of sufficient maturity.
- The estimator is asked to make a rough order of magnitude (ROM) rather than a more definitive estimate, which may subsequently become a budget. Using the assessment, the ROM would be accompanied with a low maturity score as a warning to the budget holder; it is unsuitable to use in some situations.
- An estimator may be given insufficient time to develop a more robust estimate. The assessment would result in a lower maturity level and a warning to the customer / budget holder.
- An estimator may be asked to generate an estimate with insufficient information about the scope of the project. The corresponding estimate maturity level would reflect the uncertainty in the definition of the estimate.
- The customer expects a high maturity estimate but the assessment shows why it is not possible at that time.
- The maturity level provided with the estimate can form an audit justification when the estimate is used for unintended purposes that would have required a higher maturity level.
- During a gate review the maturity level would be used to verify whether the estimate was sufficient to meet the requirements of the gate.
- The assessment could be used to check if sufficient provision has been made to take account of the uncertainty in the estimate associated with its maturity level.

4.1.4 Estimate review

The estimate review is to confirm that the estimation practices have been correctly applied and that the estimate meets an acceptable maturity, considering the information and time constraints. An independent peer review can provide guidance on how to improve the maturity of an estimate. The review can help in many ways, including:

- identifying errors in assumptions and dependencies;
- Identifying omissions in the estimate;
- identifying and minimising the effects of bias;
- identifying errors in logic;
- drawing in experts who may have different experiences;
- helping with the buy-in and approval process;
- identifying conflict with corporate strategy.

For a review to be effective, the following are recommended:

- an internal peer review is done prior to the independent review, and a maturity level is generated;
- the estimate is documented in a standard way;
- the estimate is configured and version controlled;
- the review is supported by a review checklist – prompts (see maturity assessment);
- the reviewers have domain knowledge – i.e. they can make meaningful contributions;
- the reviewers have an understanding of estimating or they have ownership of delivery;
- the lead reviewer is appropriately trained in how to conduct a review;
- the estimating team is represented at an appropriate level to answer questions;
- the review is documented, actions recorded, and actions traced to closure.

For a review to be effective, the reviewers must be independent of the estimate authors. Consider inviting some or all of the following personnel to attend during the review:

- the end customer;
- the work package owner – the person who will own the project and budgets;
- budget owners;
- stakeholders affected by the estimate – to confirm their commitment;
- finance – to confirm the financials, e.g. labour rates, cash flow, profit, etc.;
- function and resource owners – confirming their commitment of resources;
- technical specialists – to confirm the technical aspects;
- estimate process owner – to confirm estimation principles were correctly applied;
- the authors – to explain the estimate.

This is an example list, i.e. not exhaustive or definitive.

If the business is reliant on the output of a tool, without any cross checking, then the model must be subject to independent validation and verification (V&V) to the appropriate standard. The review may seek a record or certificate that a comprehensive test plan has been completed satisfactorily to assess the credibility of the cost model against the requirements of the stakeholders. This testing needs to be conducted independently by a third party and non-project staff utilising the user guide for the cost model.

4.1.5 Estimate approval

The approval is there to confirm that the estimate is ready for release and it meets the required maturity. The estimate should be signed, typically by the following people:

- the estimate author;
- the estimate reviewer;
- the estimate approver;
- the work package owner.

Depending on the value and the risk, the estimate approval may be delegated; i.e. a governance structure could be created whereby the seniority of the approvers reflects the size and risk of the estimate. For example, small and low-risk estimates can be approved by local management, while business-critical estimates are reviewed by business leaders.

5

Estimates to completion

Once the contract has started to deliver, data on the actual costs of work performed (AC or ACWP) and schedule performance will start to become available. These can be used to extrapolate and forecast the estimate to completion (ETC) this is used to calculate the estimate at completion (EAC), i.e. the EAC = ACWP + ETC (see Figure 5.1). The ETC can be derived by either a parametric or an analogous method; therefore, the good practice described throughout this guide will apply in deriving the EAC.

A parametric method will concentrate on the underlying trend and key cost drivers for changes in the trend to date and those that are expected to occur through to completion. An analogous method would be one that extrapolates through to completion based on the current earned value management cost performance index (CPI).

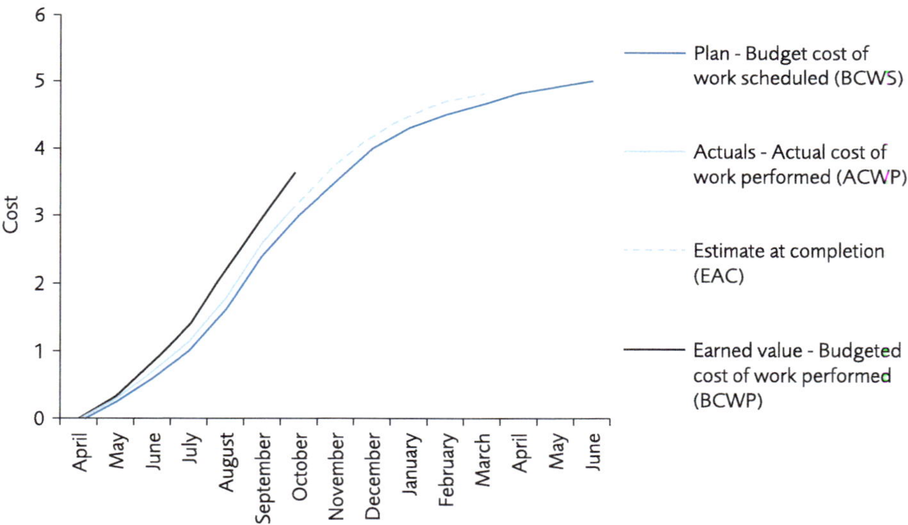

Figure 5.1 Earned value S-curve

There are numerous sources on different trend analysis techniques; these are outside the scope of this guide. See (Jones, 2019 b).

In common with the good estimating practice before contract signature, either or both a top-down and a bottom-up approach can be employed to estimate both the cost at completion and the associated date of completion. It is considered to be better practice if both approaches are used, and a sensitivity analysis is conducted to derive a 3-point estimate at completion (EAC) for each approach and method used. Ideally, these alternative perspectives should be undertaken by different parties, e.g. control account managers (CAMs) performing the bottom-up EAC for their work packages, and independent estimators creating a top-down EAC for the overall project.

5.1 EAC top-down approach

The top-down approach is normally based on a weighted average trend of the overall progress and is very good at indicating the overall EAC trend. This assumes that no major event affects the contract, i.e. risks or external changes. Earned value management (EVM) can be used to indicate an EAC, by using the relationship between budget at completion (BAC) and the cost performance index (CPI).

- An EAC derived by analogy assumes that the performance remains constant and is often based on the assumption that the latest CPI reflects the performance through to completion, so any potential variation in performance needs to be considered. This is most easily done by performing a sensitivity analysis on the drivers, e.g. using an uncertainty range around the CPI to create a 3-point estimate at completion (EAC).
- Alternatively, a parametric method may consider the underlying trend or change in the CPI over time and extrapolate that rate through to completion. A more sophisticated parametric method may consider the schedule slippage as well, and its relationship with cost using both the cost based schedule performance index, SPI and the earned schedule performance index, SPI(t). See (Jones, 2019 b).

A top-down approach is useful for testing if multiple lower level bottom-up EACs are sensible; this comparison will assist in capturing any omissions. This can be the case especially where later cost accounts have yet to start, but may be affected by good or poor performance on earlier tasks.

5.2 Bottom-up approach

The bottom-up approach to an EAC provides a useful indication of where issues may be occurring. This early warning indicator can be used to target remedial action. EVM is an example of a bottom-up approach to an EAC. Using the EVM BAC/CPI method, described above, to forecast an estimate at completion is a simplistic method in that it is fundamentally an analogical approach with linear interpolation through the origin, the assumption being that performance will remain constant at the currently observed average performance to date. The simplicity in this method lends itself well for use as an indicator, but caution should be observed in using this method to derive a high-confidence EAC. APM has comprehensive guidance on undertaking EVM so this will not be explored in detail here: *Earned Value Management Handbook* (APM, 2013).

In a similar manner to a top-down approach, a parametric method can also be used to derive an EAC for one or more cost accounts by considering either the underlying trend in the change in the cost performance, or the schedule performance, or by examining the change in any of the cost drivers used to develop the original estimate and recorded in the basis of estimate.

5.3 Risk and opportunity management in EAC

Both the top-down and bottom-up approaches tend to describe an EAC if all goes to plan; i.e. if no major risks or opportunities impact. The top-down approach may factor in some risk outturn which is dependent on the data set used to do the parametric analysis. The bottom-up approach is less likely to include any measure of risks and opportunities. This needs to be acknowledged and included within the overall 3-point EAC. Once the EAC and residual level of risk on the project are known, it is important to review the appropriateness of any outstanding contingency provision.

The methods used to evaluate the impact of risk and opportunity that were recommended as good practice prior to contract award are also considered to be good practice during contract delivery.

6

Ethics in estimation

The estimating codes of conduct lay down the behaviours we would like to see from both the estimator and the customer of those estimates. The customer is defined in this guide as the person who asks for the estimate: project manager, chief engineer, etc.

1. No individual employee, team, organisation, project (or programme), or vendor shall be required to develop, submit or certify any estimate for which they do not have appropriate confidence.
2. There shall be means to address without retribution any concerns about the integrity or ethics in the development of any estimate and those means shall be communicated clearly.
3. In order to protect the integrity, security, image and reputation of the company, senior leadership will confirm the compliance of their respective organisations to estimation policy and standards, be held accountable for the same, and shall delegate as appropriate levels of assurance and compliance to the estimation policy and standards.
4. Any known impacts to estimates, including those for remaining costs of projects in progress, shall be documented and reported as quickly as possible, and no later than in accordance with documented policy.
5. Estimate values, changes and associated impacts shall be communicated honestly, ethically and on a timely basis, to all customers, both external and internal.
6. At all times, the estimator and customer shall create an environment of mutual trust and respect. They shall provide open feedback and views without criticism. At no time shall bullying, intimidation or disrespectful behaviour be tolerated.

All professional bodies have a code of conduct which their members are expected to follow. These include, but are not limited to:

ECUK Spec – Engineering Council www.engc.org.uk/ukspec
APM – Association of Project Management www.apm.org.uk

APM – ACostE

ACostE – Association of Cost Engineers www.acoste.org.uk
CIMA – Chartered Institute of Management Accountants
www.cimaglobal.com

These cover the expected behaviour of individuals acting in a professional manner and their integrity.

Glossary

The terms in this glossary represent the views of both APMBok (APM, 2019), ACostE and the authors of the guide.

3-point estimate/ three-point estimate	[APM] An estimate in which optimistic best case, pessimistic worst case and most likely values are given. [ACostE] A three-point estimate represents three cases produced by estimating. Some organisations (and this guide) refer to these as the minimum, the most likely and the maximum. The three-point estimating technique is used for the construction of an approximate distribution representing the uncertainty of future events; this will ensure the estimate is credible.
Accuracy	The correctness of an estimate. This can be measured as the percentage error between the estimate and actual. In the case of 3-point estimates, an estimate is considered accurate if the actual cost/schedule lies inside the estimate uncertainty range.
ADORE	Assumptions, dependencies, opportunities, risks, exclusions (Shermon D, 2017)
Assumptions	A statement that is taken as being true for the purposes of estimating, but which could change later. An assumption is made where some data is not available or facts are not yet known.
Baseline	The reference levels against which a project, programme or portfolio is monitored and controlled.
Bottom-up estimating	[APM] An estimating technique that uses detailed specifications to estimate time and cost for each product or activity. Also known as analytical estimating. *This should not be confused with the 'Estimating Method of Estimating by Analogy' (Section 2.3.1).* [ACostE] An approach to estimating all individual work packages or activities with appropriate level of detail, which are then rolled up to higher-level estimates. The accuracy of bottom-up estimating is improved when individual work packages or activities are defined in more detail. *See section 2.2.2.*

Glossary

Budget	A budget is the allocation of funds and resources to a project. Budgets should normally fall within the range of the 3-point estimate. The budget may reflect a perceived affordability level, which is outside the estimated range.
Comparative estimating	An estimating technique based on the comparison with, and factoring from, the cost of similar, previous work. Also known as analagous estimating.
Confidence (statistical)	A confidence level is an expectation of percentage probability that the outturn value will be less than, or equal to, the specified value. A confidence interval is the difference between two confidence levels. They do not have to be symmetrical; e.g. 90 per cent to 30 per cent.
Confidence (management assurance)	The confidence in the estimate is provided to management through the application of multiple estimating methods within a repeatable process.
Contingency	[APM] Provision of additional time or money to deal with the occurrence of risks should they occur. [ACostE] Some companies may also decide to use contingency to allow for variable performance (uncertainty).
Cost breakdown structure (CBS)	The hierarchical relationship between the labour and non-labour resources (e.g. materials) in a work package, which would be aligned with an organisation's accounting systems.
Cost rates	The chargeable unit of throughput or output of resource consumed, e.g. labour or equipment. Often charged by the hour or day.
Costs	Costs are defined as the internal impact to a business. Cost is not the same as price. Price refers to the addition of financial considerations, such as profit and warranty, and is usually offered to the customer or client.
Dependencies	Equipment, data, information, assets or items required from third parties with an associated scope. These dependencies are assumed to occur in order for the estimate to be valid.
Estimate	[APM] A forecast of the probable time or cost of completing work. [ACostE] A numerical expression of an approximate value, expected to occur, based on a given scope, defined by recorded parameters and assumptions. Usually associated with cost and schedule durations, but it can also refer to other entities, e.g. technology readiness levels.

Estimate at completion (EAC)	The estimated cost or time of the project forecasted to the time of project completion. This can also be referred to as forecast at completion (FAC).
Estimate to completion (ETC)	The difference between estimate at completion (EAC) and the current situation. Can be expressed as either cost or time.
Escalation rates	Escalation is the inflation factor used to take into account the changes in cost due to the financial economy over time. It is good practice to show the inflation factor applied to each year.
Estimate purpose	A statement of what the estimate is to be used for. This will define the methods, documentation, precision and effort required to generate the estimate.
Estimate uncertainty	*See Uncertainty.*
Ethereal approach	The act of accepting a value without the full understanding of how it has been derived. *See section 2.2.3 Relying on the work of others – 'ethereal' approach.*
Expert judgement (subjective)	Use of knowledge gained from past project experience to express an informed opinion, in the absence of definitive data.
Exclusion	Work content which may or may not be delivered, but is outside the scope of the estimate and is stated for clarity.
Forecast	A prediction of a defined future state, typically related to the duration and out-turn cost of a project or programme.
Inclusion	Processes, deliverables, products, materials and services that are within the scope of the estimate.
Maturity	An expression of the level and clarity of definition of scope, and availability of suitable comparable data from which the estimate has been created.
Mitigation	*See Risk*
Monte Carlo analysis	A probabilistic method for simulating the likelihood of potential project outcomes.
Normalisation	Normalisation takes the actual cost of previous, similar projects as a baseline and then adjusts for known differences, such as size, complexity, scope, and duration.

Opportunity	[APM] A positive risk event that, if it occurs, will have an upside/ beneficial effect on the achievement of one or more objectives. [ACostE] The potential for a situation that may improve the project's outcome, performance, cost and/or schedule. A beneficial effect or event that may or may not happen, with an associated probability of occurrence less than 100 per cent. Opportunity promotion is the action of increasing the probability or impact of an opportunity. If promotion is possible and approved, then the post-promoted opportunity is included in the estimate and the promotion costs and activities are planned into the estimate. An opportunity can be treated in the same way as a risk, however some organisations refer to 'risks' as being only negative events, whilst opportunities are positive events.
Parametric estimating	A systematic means of establishing and exploiting a pattern of historical behaviour between the variable that we want to estimate, and one or more other independent variables.
Product breakdown structure	The hierarchical relationship between the components and sub-systems that comprise the system.
Precision	The precision refers to the tolerance of a value, expressed as a +/- spread, defined in terms of an absolute or percentage value. This is often referred to as uncertainty.
RACI matrix	Responsible, accountable, consulted, informed – a management framework for managing actions, in terms of responsibility, accountability, and how stakeholders are consulted and informed.
Responsibility assignment matrix (RAM)	A diagram or chart showing assigned responsibilities for elements of work. It is created by combining the work breakdown structure with the organisational breakdown structure – (APM, 2019)
Requirements	The stakeholders' wants and needs clearly defined with acceptance criteria.
Request for information (RFI)	An RFI is primarily used to gather information to help make a decision on what steps to take next. RFIs are therefore seldom the final stage and are instead often used in combination with the following: request for proposal (RFP), request for tender (RFT), and request for quotation (RFQ).
Request for proposal (RFP)	An RFP is a solicitation made, often through a bidding process, by an agency or company interested in procurement of a commodity, service or asset, to potential suppliers to submit business proposals. It is submitted early in the procurement cycle, either at the preliminary study, or at the procurement stage.

Request for quotation (RFQ)	An RFQ is a standard business process whose purpose is to invite suppliers to bid on specific products or services. An RFQ typically involves more than the price per item. Information like payment terms, quality level per item or contract length may be requested during the bidding process.
Risk	[APM] The potential of situation or event to impact on the achievement of specific objectives. [ACostE] It is acknowledged that in many industries and organisations the term 'risk' is used exclusively for negative impact events, whereas the term 'opportunity' is used for positive impact events. Where organisations use 'risk' as a term which has positive or negative impact, the term 'threat' is used to denote negative impact risks. In both cases, risks may or may not happen with an associated probability of occurrence less than 100 per cent. Risk mitigation is the action of reducing the probability, severity, seriousness, or impact of a risk. If mitigation is possible and approved, then the post-mitigated risk is included in the estimate and the mitigation costs and activities are planned into the estimate. A risk can be treated in the same way as an opportunity.
ROM	Rough order of magnitude – representing a ball-park estimate. However, the exact definition of what precision (+/-) it represents is unclear. Whatever bands you choose, recognise that such estimates are for information only. The range is pessimistically skewed e.g. −10 per cent/+40 per cent.
Threat	[APM] Recognises a threat as a negative risk event. [ACostE] Recognises that some organisations consider all risks to be negative – *see Risk*
Target	The term target is used in this guide to mean the goals of the project. In most cases the targets will be cost and/or schedule but may also include maturity goals, resource limitations etc.
Top-down	A high level estimate based on top level project assumptions without the need to understand all of the project detail. Top-down may cascade down to lower level by applying the same principle to the next level down in the work breakdown structure. *See section 2.2.1*

Uncertainty	Uncertainty is the inherent and potentially uncontrollable variability in estimating the cost and schedule. This can be due to a number of variables, e.g. poorly defined scope and variable historical performance. Uncertainties arise because the organisation cannot (or does not) have a complete understanding or control of the variables. An uncertainty is something that will happen but the effect on the project outturn cannot be known precisely. However, the effect will be expected to lie within a defined range.
Work breakdown structure	The hierarchical relationship between the functional work packages. It includes all the elements for the hardware, software, data or services that are in the scope of the project. It links the product breakdown structure (PBS) and the organisational breakdown structure (OBS).

Acronyms and abbreviations

ACWP	Actual cost of work performed
ADORE	Assumptions, dependencies, opportunities, risks, exclusions
BAC	Budget at completion
BCWP	Budget cost of work planned
BCWS	Budget cost of work scheduled
BOE	Basis of estimate
CAM	Control account manager
CBS	Cost breakdown structure
CPI	Cost performance index
EAC	Estimate at completion
ETC	Estimate to completion
EVM	Earned value management
FAC	Forecast at completion
MDAL	Master data assumptions list
OBS	Organisational breakdown structure
PBS	Product breakdown structure
PESTLE	Political, economic, social, technological, legal and environmental
RACI	Responsible, accountable, consult and inform
RAM	Responsibility assignment matrix
RFI	Request for information
RFP	Request for proposal
RFQ	Request for quotation
ROM	Rough order of magnitude
SBS	Service breakdown structure
SME	Subject matter expert
SPI	Schedule performance index
SPI(t)	Earned schedule performance index
WBS	Work breakdown structure

References

APM, 2019. *APM Body of Knowledge 7th edition.* Princes Risborough: Association for Project Management.

APM, 2013. *Earned Value Management Handbook.* Princes Risborough: Association for Project Management.

Comptroller and Auditor General, 2001. *Modernising construction, HC 87 Session 2000–2001*, London: National Audit Office.

Comptroller and Auditor General, January 2017. *Forecasting in government to achieve value for money, HC 969 Session 2013–14*, London: National Audit Office.

Flyvbjerg, B., 2003. *Megaprojects and Risk: An Anatomy of Ambition*, s.l.: Cambridge University Press.

HM Treasury, 2018. *The Green Book, Central Government Guidance on Appraisal.* London: Crown Copyright.

Jones, A. R., 2019 a. *Working Guides to Estimating and Forecasting: Volume I – Principles, Process and Practice of Professional Number Juggling.* Abingdon: Routledge.

Jones, A. R., 2019 b. *Working Guides to Estimating and Forecasting: Volume III – Best Fit Lines and Curves, and Some Mathe-Magical Transformations.* Abingdon: Routledge.

Jones, A. R., 2019 c. *Working Guides to Estimating and Forecasting: Volume V – Risk, Opportunity, Uncertainty and Other Random Models.* Abingdon: Routledge.

KPMG, 2015. *Climbing the curve, 2015 Global Construction Survey*, s.l.: s.n.

Labaree, L. W., 1961. *The Papers of Benjamin Franklin, Vol. 3, January 1, 1745, through June 30, 1750.* New Haven: Yale University Press, pp. 304–308.

Legislation.gov.uk, 2015. *Data Protection Act 1998.* s.l.:s.n.

National Audit Office, December 2013. *Over-optimism in government projects*, London: National Audit Office.

National Audit Office, 2018. *NAO survival guide to challenging costs in major projects*, London: National Audit Office.

Norden, P., 1963. 'Useful tools for project management'. *in Dean, BV (Ed.) Operations Research in Research and Development,.*

Shermon D., G. M., 2017. *Cost Engineering Health Check*. s.l.:Routledge.

Shermon, D., 2017. *Uncertainty and Risk: defined through probability and impact.* [Online] Available at: https://www.linkedin.com/pulse/uncertainty-risk-defined-through-probability-impact-dale-shermon/

Smith, E., 2013. *Estimate Maturity Assessments.* Sandbach, Association of Cost Engineers.

Turré, G., 2006. 'Plant capacity and load'. *in Foussier, P, Product Description to Cost: A Practical Approach,*, Volume 1: The Parametric Approach, pp. 141–143.

ND - #0164 - 080726 - C68 - 246/189/4 - PB - 9781903494844 - Gloss Lamination